All images in this book are the authors or NASA Veimages or Composite images made by author over NASA Veimages to show the lay of the land.

Thanks to NASA for there scientific endeavors that help the studies of our planet.

Image from The Visible Earth @
http://visibleearth.nasa.gov/

This book is dedicated to
My wife Dina who spotted the
Sleeping Fox
Menno & Cara Hershberger
For there friendship and believing
In me

to Copernicus for writing the Heliocentric
theory about the solar system and Galileo for
speaking out in favor, even when he new that he
would be jailed for speaking out. Sometimes stories
have to be told.

Table of Contents
Chapters in this book are divided into the Richter scale

Introduction to the Sleeping Fox

From the beginning of time the earth has seen many changes in it's geologic path to the present. There have been numerous extinctions due to catastrophic events that have altered the land, the atmosphere and oceans to get to where we are today. Due to the rapid development of science we have probably learned more in the last 50 years than we have ever known through out the history of mankind. And even then a lot of what we know today is in theory. But if we had not had the earth shattering events that buckled the earths crust we would probably know a lot less because it has enabled us to see what is underneath by bringing it to the top.

In the following pages I will try to bring to life a little sliver of geologic time that changed the earth and brought us a great deal of our knowledge through one catastrophic event and gave us a Sleeping Fox buried in the ancient rocks of earth.

As human's we have always thought of our world as being the place where everything has begun and ended, and for us it has. We have always thought of everywhere else as outer space but in fact we ourselves are in the vast open never ending void of this place we call space. But as people we have become comfortable within this space we call home just as we would in our own home, on our own block, town, state, country and yes our own space.

When something or someone unfamiliar come's into this space we may become concerned or defensive, so sometime's it may be hard to understand something different when we have become comfortable with our past the way we have been taught.

In the following pages I will try to paint a very different picture of our western U.S. (that to some) may become offensive. I have heard it said that a picture is worth a thousand word's so I'll show you the picture's and tell you my theory and let you be your own judge of what you see but I know it will change the way you see the world everyday for the rest of your life because you will see it every time you look at a map in a book or on TV.

So this is my theory about the way thing's may have been in the past and how a sleeping fox was discovered in the west.

By now we have all heard more story's and theory's about comet's, meteor's and asteroid's than we care to remember but here is one that has some very real and very interesting detail's about a certain geographical area in the western United State's that will make you look twice every time you see it.

I, like so many of us out there have always gazed up at the night sky in wonder and amazement at what was out there or just took in the beauty of it all, the moon so big and bright, the billion's of star's and those little one's that fall from the sky but then there are those of us that after the beauty of it we want to know why and how?

I believe that everything on our planet has came from somewhere else in the solar system and has come together in just the right way, mixing all the best ingredient's necessary for life. By studying the comets and meteorite's through science it's been determined that they are made up of different ingredients. Comets are mostly ice and meteorite's can be made up of different metals or softer materials and sometime's just loose material's held together by a small amount of gravity. In recent scientific discoveries there have been life forms found in the most inhospitable place's known to man so therefore it's likely that they have with stood the perils of space travel.

In the billion's of year's that our planet has been evolving it has probably been hit million's of time's by meteorite's and numerous time's by comet's that have cooled our planet to the state that it is in today. The meteorite's have more than likely brought us our precious metal's that we have by plunging through the earth's crust and then being melted by the intense heat and then been redistributed by the magma through volcanic activity. And the water and the seed's of all life may have been brought here by the way of the comet.

Year's ago I heard that the crater's on the moon were ancient volcano's but for some reason that just didn't set well with me because they didn't resemble volcano's here on earth. Then I heard a scientist on T.V. talk about impact crater's and I believed he was on the right track because to me it just fit like the piece's of a puzzle. And then as the scientist's watched the comet hit Jupiter there was a whole new chapter beginning in astrogeology and that is when I think I got hooked on studying the earth through satellite and weather image's.

I have always been somewhat been interested in the weather taking photo's of cloud's and lightning and watching the weather on T.V. About the same time I was getting into the impact site theory I saw something on the weather satellite image's that amazed me, right in front of my eye's was a perfect circle in the cloud formation's following the lay of the land. After I saw this unique feature I was hooked even more and started studying the weather satellite image's everyday to see if I could catch this again but to my surprise it didn't happen very often but you can see this area with out the aid of the clouds easy. In fact once you pick it out you will always see it.

I am lucky enough to have traveled through this area over the last thirty five year's or so to be able to study it (on a non professional side though) to see enough that what I'm about to tell you is mostly true.

Over the year's of studying this I have e-mailed a few scientist's and other people with out giving out to much information to see if anybody was interested but no one appears to be, so that is why I started to write this book., So I hope you find it interesting, maybe educational or if nothing else it may stir up some debate among the professional scientist's.

page 4

The area of which I speak is the four corner's section of the U.S. When I spotted the area in question, there was a hole in the cloud's in the perfect shape of an impact crater, round and with a point of clouds in the very center just like a bull's-eye. There were two other sites to the north that you could see also, in northern Utah and in Wyoming. After seeing this I watched and waited for it to happen again. Over the years of watching the weather I have heard the forecaster's say that a low or high pressure system was caught in the four corner's area and struggled to get pushed on, and now I know why.

In cross section's of the western state's from the front range of the Rockies' to the Great Basin area, from the bookcliff area of eastern Utah to the central part of New Mexico and Arizona you can see that it has the shape of a bunt cake bowl (as per the shape of an impact crater with a mound of residual debris in the middle) that is approximately 650 mile's long and 500 mile's wide in an oval shape.

There are three more to the north that are impact site's as though it was a multiple impact (as the one that hit Jupiter) in northern Utah around the Vernal area which is referred to as the basin and in Wyoming that take's in most of the central and southern part of the state and a fourth in the Riverton to Cody, Wyoming area.

The area also has a unique feature about it as it has the appearance of a sleeping fox. As I studied the area as an impact site my wife said to me one day that it look's like a fox, and there it was bigger than life right in front of me ,so I have taken the liberty of naming it the Branham's Sleeping Fox Comet impact site.

This impact site is probably the largest site of its kind visible on earth and will be the most visible land feature from space for year's to come.

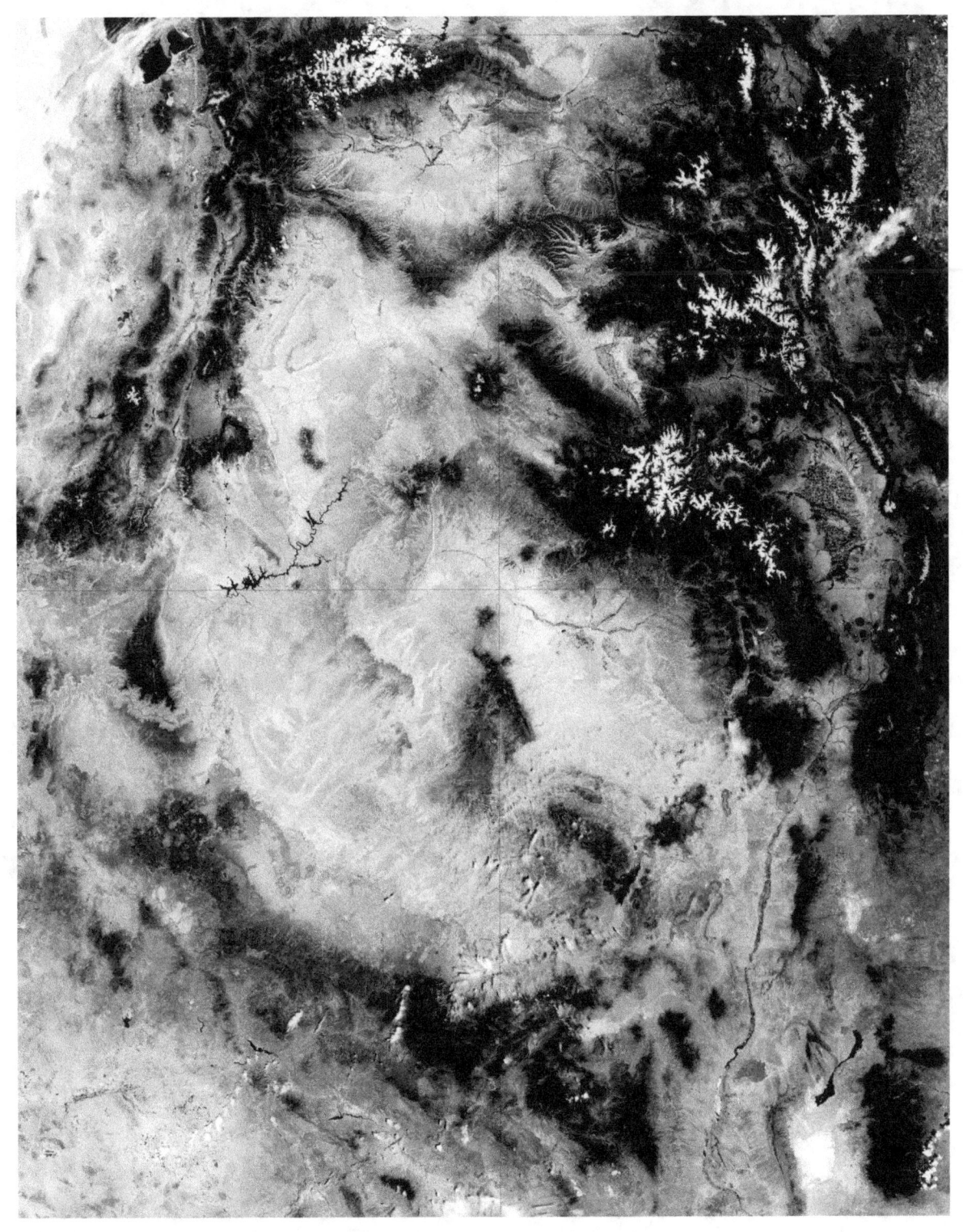

Sleeping Fox area not outlined.
The Four Corners area of the southwest.

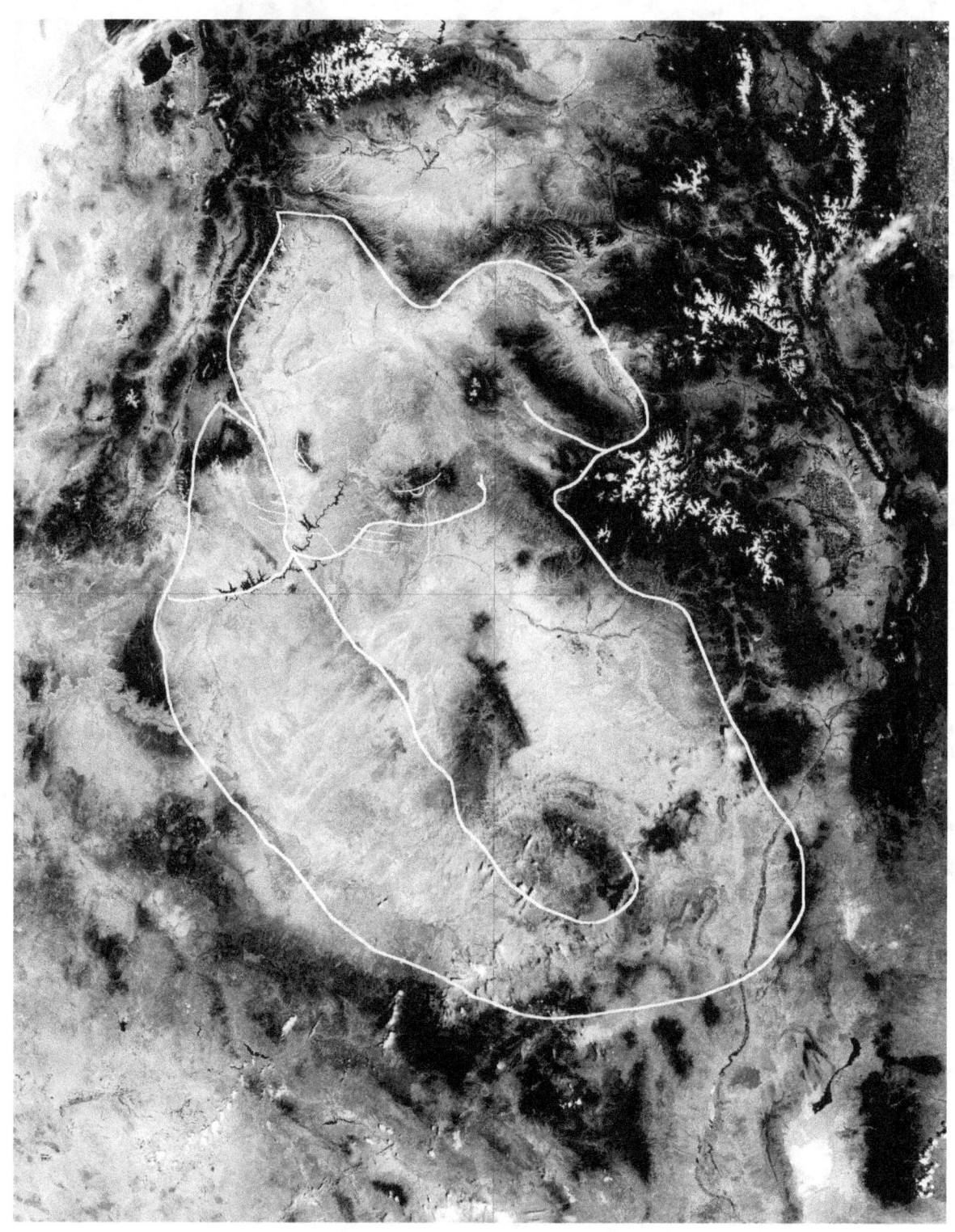

Sleeping Fox outlined

Imagine around the four corner's area the land being of low elevation to the point that it was just above sea level. There were vast inland sea's of fresh water and salty mixed water's where prehistoric ocean's and fresh water's met. There were great island's of lush vegetation with huge tree's where there was very little or no change in season's and plant's and animal's had little or no worry of starvation due to annual climatic change's, where the only thing they had to worry about were the frequent hurricane's and water spout's that came up through the gulf of California that stretched up to the Great Basin area or the Gulf of Mexico that came as far north as southern Utah and western Colorado.

We know that the mountain's of the west capture the moisture from the Pacific and dump's it on the top's and leave very little for the valley's and then when it get's to the mid-west it reorganize's to dump large amount's. Before the mountains were large and tall and there was nothing to get in the way, the moisture laden clouds could drop the rain's in the west creating vast rain forests through out the western part of the continent.

page 7

In the area we now know as the four corners area of the southwestern part of our country there was a great oasis of green and blue. Great creatures swam in the vast nutrient rich water's and they fed on lush green plants of the river delta's and the giant trees. There were the great fish of the water, the plant eater's, the meat eater's that fed on the dead as well as the living.

As the great creature's went about there daily routine's as they have for thousand's of year's they didn't know that thing's were about to change forever.

As night fall's over the delta the creature's are oblivious to the little light that has appeared in the night sky over the horizon. As time goes on the light grows larger and larger until it start's to light up the night and disrupts the animal's sleep and feeding habit's. Coming from the opposite side of the earth as the sun the light from this comet is was so bright at night it was as if there is two day's in one. Some of the animals take advantage of this by feeding for longer hour's which in turn take's a great toll on the others by stressing them to the brink of exhaustion. As the inevitable happens the comet slam's into the earth shaking it to the core and causing a shockwave to radiate out taking down the animals and vegetation across the whole continent.

Comets being made up of mostly ice wouldn't have a solid impact such as an iron meteorite, so therefore it wouldn't create a fireball or explosion like one would see in the Hollywood movies or T.V. shows. The light given off by a comet is mostly reflected light so as it entered earths atmosphere it would start to vaporize and be more like a cloud, but a comet the size of the one that hit the earth at the four corners would have hit the earth on one side before the other side came into our atmosphere as the impact site is about 600 mile's long just in one area and there are at least three sites.

A comet floating through space would be like a ball of gelatin picking up anything that came into contact with it ,that wasn't so large it couldn't catch and hold. But it could catch matter such as sand and fine particle's of other planets that has been dislodged by meteor's hitting other heavenly body's including life forms. Page 9

When the comet hit the earth almost directly on the four corner's area it came in from the south southeast to the west Northwest and as it hit the earth lurched to the north and gave us the wobble in the spin of the earth that gives us our four seasons that we have today and probably slowed the spin slightly . Hitting the earth at this angle and speed in three or four areas it pushed the crust to its breaking point and shoved the Rocky Mountain's up almost vertical, but caused a ripple effect to the west by lifting the crust and cracking in north to south lines, but radiating out from the center of impact. As the earths crust rose and got pushed to the west it caused the water from all the inland sea's and water way's that covered the Great Basin area to drain out to the south and over time dry up leaving behind the remains' of the sea animal's shells. And as the earth's crust broke it opened up causing thousand's of small volcanoes to pop up through the crack's across the west. As you look at the topography of the west from north to south most all major valley's run north to south, that is because of the way the earth fractured from the impact. If you look at the way ice buckle's on a large body of water you can see what I mean.

In the southern portion of the impact site you have the Mogollon Rim near Flagstaff, Arizona which is where they do a great deal of studying of astrogeology. And to the north of there we have the Kaibab Plateau that rises up to create another part of the Impact site.

And then there is the most well known part of the and that is the Grand Canyon which is where I believe most of the water

From the comet drained out to sea. The ice from the comet covered most of the western U.S. if not all. But the majority stayed within the site and as it melted it had to go somewhere. When the earth buckled and rose a large crack formed at the site of the canyon and over thousand's of years as the ice melted it gave us something Grand.

To the north there is the basin area around Vernal, Utah that is the second but much smaller site. And to the north of that there is southern Wyoming area from Evanston to Laramie and around Riverton to Cody, Wyoming there is another site.

When you travel through this area it may look as though they are separate areas but from an orbiting satellite it looks as one. But over the millennium as the wind blew and the rain's washed the

mountain's down and they have filled in the valleys. It has changed a lot, but there are still area's of evidence that can be used to get to the bottom of it all. Take for instance the Water Pocket fold in south eastern Utah. On the east side of the fold as you drive west there is a layer of earth that is laying perfectly flat at the bottom and just a few hundred feet to the west the earth is almost vertical and the soil that is there look's as though it has been tilted up and then washed down. The rock formations have been tilted up and you can see the different layer's from being pushed down and over at the same time and farther to the west the ground just rose up in a mound. Because of the great stress at the east edge the ground fractured.

Lake Bonneville is the big brother of the Great Salt Lake which used to cover the Great Basin area and may have been drained by an impact. The lake was vast and being fed by fresh   - Page 11 water river's.

So with the great upheaval and the climatic change the lake dried to its present state which raised its salinity

At different times during it's life the giant lake had multiple levels of water due to it's outlet being blocked by land and ice dams. At the time of the comet impact the land rose and was covered wih ice. Over time as the ice melted the lake filled to great depths there was another earth shattering event (this will be explained later in the book) that broke a great ice dam and released vast amounts of water until the water level reached the land blockage.

After this happened the water level stayed stabel until due to the ice still melting the great amounts of water washed away the earthen dam and the lake drained even lower until the ice was depleted and evaporation over took the rate of fill.

Lake Missoula was also a great body of water held back by an ice dam that broke and released vast amounts of water down the Columbia River Basin. With the melt water from the Yellowstone area, Lake Missoula and Lake Bonneville the amount of water that flowed down the Snake River and the Columbia River would be another great event of geologic proportion.

Both lakes existed and drained about the same period of time, so it makes sense to believe that they were shaken by a catastrophic event and there waters released.

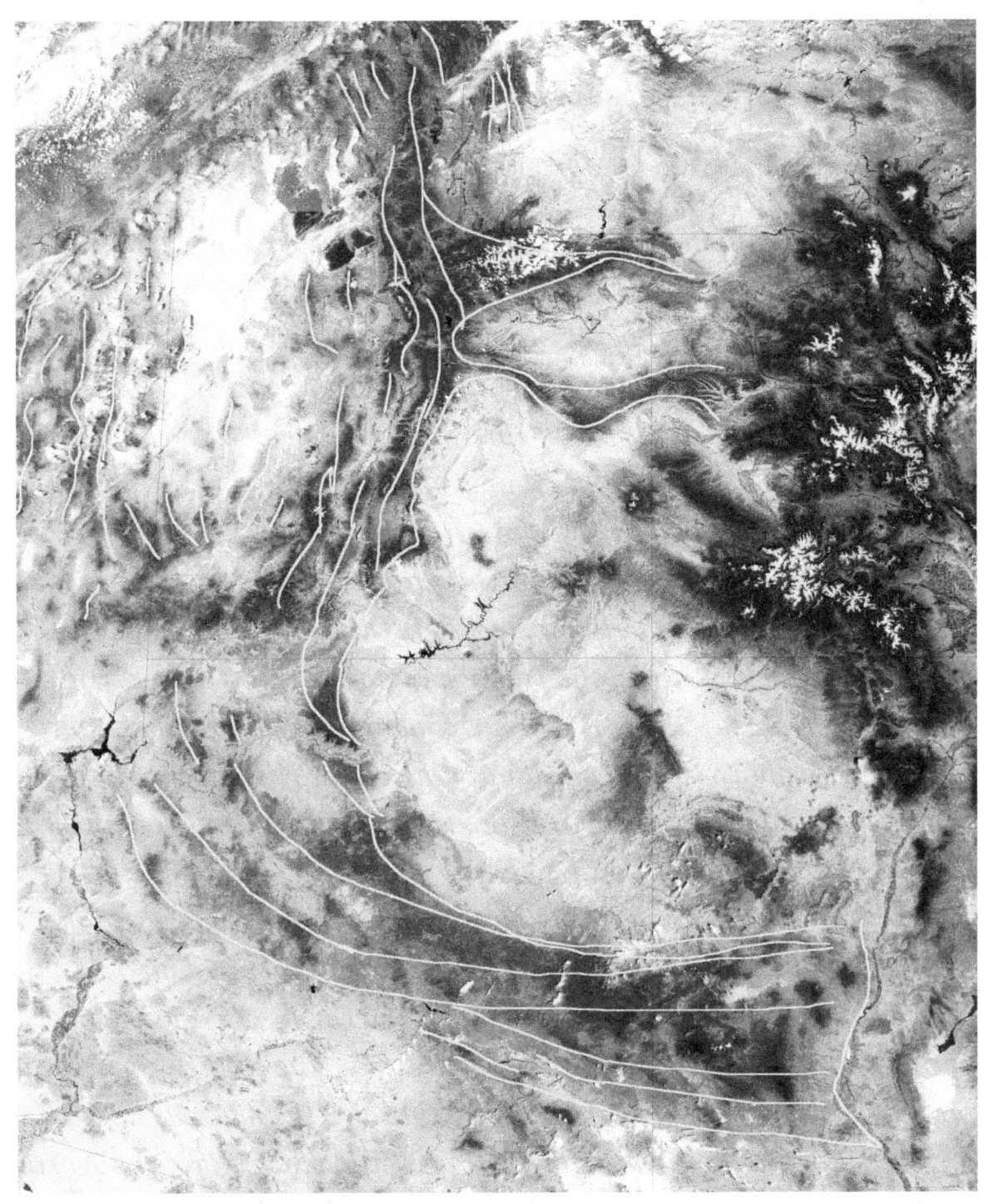

In this photo I have outlined the mountain ridges to show how they radiate out from the impact site, from the Mogollon Rim throughout the Great Basin.

After millions of years of silt and mud from the weather and volcanoes and impacts of other worldly objects which made layers of sediment at the bottom the shallow sea's the earth was ready for another chapter in history. As the layers got deeper the sediment below started to solidify from the weight above. This is why in some places as in the Waterpocket Fold it seems as though the top layer is softer because this was the top layer of mud in the shallow sea. Of course this would depend on the type of sediment and conditions, so this may not be the case in all area's.

In all of the valleys in the Great Basin area were the mountains capture sufficient moisture there is always underground water caught in the underlying area's of sediment that has filled the voids that were created by the fracturing and tilting of the earths crust and sometimes the water flows underground for long distances and will rise above the ground for awhile only to disappear back into the ground at another place due to the depth and width of the sediment between the fractures.

As the compression waves traveled through the west and created a catastrophic convulsion of the earths crust like no other, it came to where the Sierras are today and the earths crust broke in one giant slab from the lift of the earths crust due to the downward pressure at the impact site on the east side and the massive upward pressure (due to the subduction of the Pacific plate under the North American plate) at the crustal boundary, the North American plate broke and the Pacific plate shoved it upward and held it in place. When the great slab that we now call the Sierras dropped on the west side due to it's enormous weight and the Pacific plate pushed it to the east a little it left a great void that we now call the San

In this photo you can see how the earth broke from the downward force and was pushed up in a large dome to the west from the mass of the comet and the east to west motion of the impact.

In this photograph you can see that the layers of rock and soil on the left lay perfectly flat, while the large jagged rock layer on the right stands almost vertical.

San Rafael Swell area in eastern Utah, with I-70 in the foreground

Joaquin Valley. As the earth tilted up massive amounts of plant and animal matter washed down into the valley as well as salt water from the Pacific ocean and after millions of years of sediment it filled and became the valley we know today.

From the ice of the comet and the resulting climate change the Sierras was probably covered in glaciers like the one that carved Half Dome in Yosemite National park.

Due to the massive weight of the great Sierra slab pushing against the Pacific plate and the plate trying to push under and north it is no wonder that there are so many earthquakes in the great state of California.

In the future, when the pressure from the North American plate pushes west due to the weight from the comet debris and rise in elevation, the Pacific plate shoves it self under the great Sierra slab of granite to the point of fracture, it may slip down into the vast depths of the earth only to be reborn millions of years from now. If this does happen, then Reno, Nevada may become the new beach front property.

This would also cause a great tidal wave would devistate the Eastern World and would again shake the earth to it's core.

Scientist's talk of great upheavals and fault lines but no one seem's to agree on how they got there and when. I believe that the comet theory answers a lot of those question's because of the way the mountains and the fault lines run north to south.

The Rocky Mountain's and the Sierra's were most likely already here but at a lot less height than what they are today. The fault lines in the west are likely to be the product of the cracks and pressure ridge's due to the fracturing and buckling of the earth's crust. When the comet hit the earth it had so much debris that it raised the ground elevation several thousand feet due to the enormous size and is probably still buckling to this day by way of outward pressure and the subduction of the Pacific plate under the north American continent.

The massive weight of the continent to the east would prevent much of the pressure from going east but where the tectonic plates meet to the west and move it could let the earth move ever so slightly just enough to let there be small amount of crust fracturing throughout the Great Basin area fault lines.

Who knows maybe the great National Parks and the great sand deserts of the west are left over remnants of the comet and that is why the area is so barren of vegetation. The sandstone that make's our beautiful park's and canyon's of the west may just be what's left over after the comet melted and washed away to the sea through the Grand Canyon.

I also believe that it may have played a part in the formation of the Petrified Forest. If there was a great fireball as some would believe it would have burnt the tree's and just left ashes.   When the comet coming in caused a pressure wave of heated steam the fine particles of mineral's being of different consistency may have softened the fibrous material's of the tree and fused themselves into the shape of the tree's. If we look at the Petrified Forest the trees are laying on top of the ground and close to the edge of the impact area. We all know that there are fossilized tree and fern's in the ancient mud of time, but only impressions are made of these.

These petrified tree's have lain buried under what may be the remnant's of the comet or the sand's of the many lake's and sea's that were redeposited by the comet, until the wind's of time has blown them away and washed them to the southwest through the Grand Canyon and down to the great desert's of southern Arizona and California where they are still moving to this day.

page 16

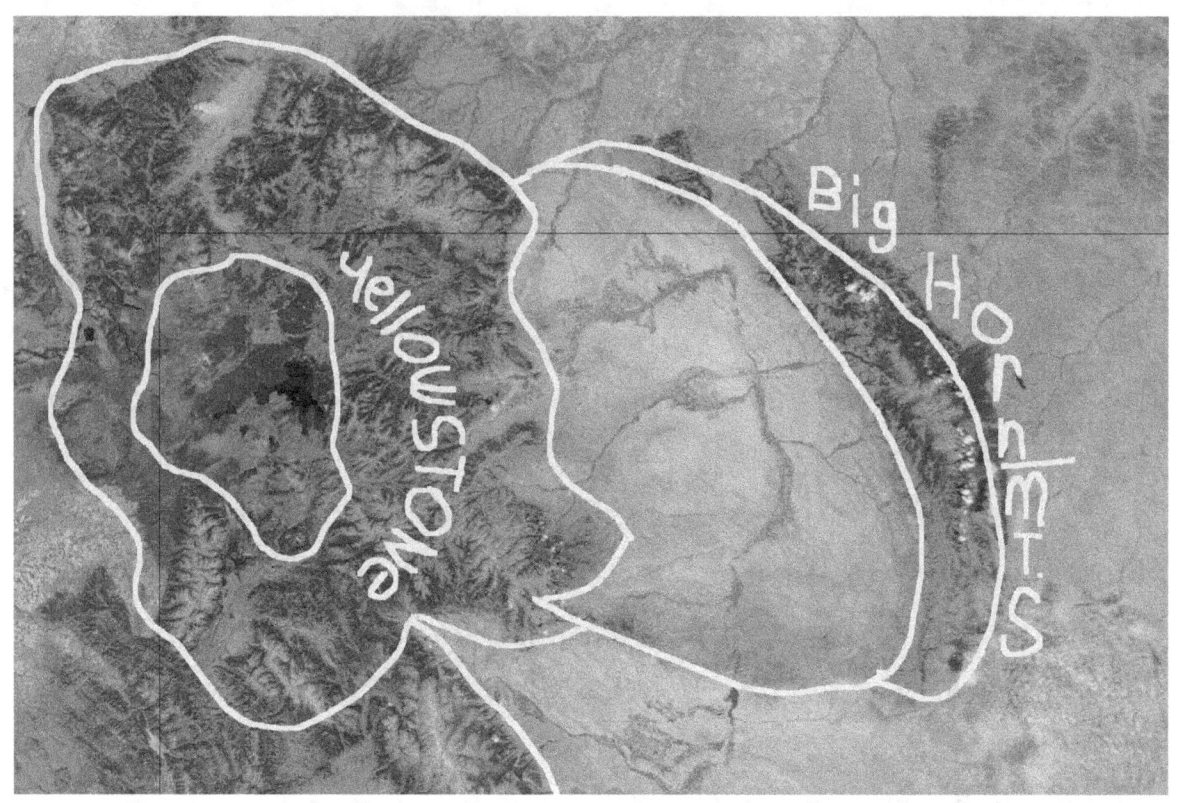

In this photograph you can see the Big Horn mountains in a half circle on the northern end of the impact site and Yellowstone on the left in more of a full circle. The reason for this is because the meteorite that hit Yellowstone hit at a later date in an almost straight on impact. The earth was already fractured therefore it penetrated the crust causing vertical uplift to the surrounding area. If you look at the mountains that surround Yellowstone they are young having hardly any wear. Yellowstone is still trying to heal this wound.

In this photograph at Flaming Gorge you can see how
the earth buckled and cracked, dropping on one side to
create the half circle on the right side. You can see the
different layers of sediment that was laid down in the
millions of years before. When the comet hit, the earth
below revealed it's self to the world.

As I studied the northern part of the impact site , I could tell that there is a mixture of geologic events there. As you look at the Big Horn Mountain's of northeastern Wyoming, you can see that they have the crescent moon shape that would be evidence of an impact from the east, southeast as the others were. Pushing up the east side almost vertical, and pushing the other side to the west.

If you take a good look at satellite photos you can see that all of the land within these areas appears to be of the same material composition, and lack of vegetation. Now some scientist's would probably say that it is due to lack of moisture, but I think the evidence show's otherwise, except that we do know that the mountains tend to capture the moisture. So I believe that the lack of water has preserved the evidence.

Now, I am going to go out on a limb here, but this is my theory about Yellowstone National Park. The scientist's tend to believe that Yellowstone is a mega volcano, but by looking at these satellite image's it appears to be another impact site to me. And by now I am sure everyone is thinking oh no not another one, but hear me out at least.

I don't believe that it is part of the same scenario, but at another time altogether. It look's as if a large solid meteorite struck the area at an angle just slightly from the west southwest but almost straight on. It hit with such a force that it penetrated the earths crust in an already vulnerable state from the comet impact. Because the earths crust was already broken, the meteorite penetrated with relative ease. By plunging through easily, you would not have the wide spread devastation that come's with a solid surface impact. But you would have the outer perimeter lift that would circle the impact.

So that you can understand why the difference in a solid and a soft impact is not the same, I will try to explain the best I can.

If you take a snowball and throw it at the ground at an angle, the weight of the snowball will hit with the force at the same angle at which it is traveling and then explode in that same direction leaving very little of that energy to transfer in the opposite direction. Now, if you take a heavy solid object and throw it at the ground at the same angle it will penetrate the ground and the energy will force the ground to move in the same direction up and away from the object leaving a mound equal to the force and mass of the object, but not much in the opposite direction. Now if the object come's straight in the force is equal in all directions.

When the meteorite hit it blew out very little of the surrounding area but caused it to lift because of the force and the mass. After the meteorite entered the crust it left a big hole that filled with lava and eventually cooled and formed a cap that most scientists believe is a lava dome as in a volcano. I am sure that over the eon's of time there has been events where the crater has let off steam because of the subduction of the Pacific plate under the north American continent but I don't believe that it has blown out like a mega volcano. Because of the massive amount of lava that it would take to fill the opening in the crust it may never heal the wound. And with all of the volcanic activity along the Pacific rim letting off pressure I believe we are probably safe for now from a mega event from the Yellowstone caldron.   Page 18

Lava domes such as this exist throughout the western U. S. and most of which are in between the mountain ranges or at the base where the land buckled and let the magma flow up from the cracks. The earth heaved it's self to the west and opened up but when it reached the spot where the Great Sierra Slab is the earths crust pushed the Sierra's up and the weight of the Pacific Ocean and Pacific Plate shoved back the North American plate pushing the cracks back together and lifting the Sierra's up to where they are. This caused the west to become dry and arid.

In this photograph, these peaks are lava domes protruding out of the almost perfectly flat ground in a straight line almost due north. Following a crack in the earths crust and due to the liquifaction of the loose soil around it that had been a large body of water, it settled into a large open barren flat .

Because of the rise in elevation after the impact, the resulting mountains capture a lot of rain and snow that has washed those same mountains down into the hole it created. But because it has not completely cooled yet, we still get the shifting and rise and fall of the Yellowstone Lake area and all the gases that come out of the ground like in a volcano.

In and around the Yellowstone area (southern Montana, eastern Idaho and northwestern Wyoming) there are numerous dormant volcano's that were probably made as a result of the impact of both the comet and the meteorite when the ground fractured.

By looking at other volcanic activities, it makes sense to say that a mega volcano would leave a mound of rock and ash at least a thousand times larger than Mount St. Helens, that is a big pile of rock! Where has it gone? There is no doubt that there have been many expulsions of rock and magma in the area. When you look at the surrounding mountains they stand tall and the valleys are deep, they are bare of the mountains of rock and debris that a mega volcano would produce, but not free of the smaller amounts of lava and ash that would come with the activity of fractures.

With all this volcanic activity over hundreds of years there would be enough ash in the air to pile up to the east burying the landscape and killing off the land animals of the day. The large amount of ice and snow created by the comet would have surely washed most of the debris left by the comet and the meteorite, to the sea at first by way of the Snake River Canyon just as it did at the Grand Canyon at various rates of melt depending on the volcanic activity. Page 19

When the comet crashed into the earth it covered most of the western U.S. with hundreds if not thousands of feet of ice, and most likely the whole earth was affected to some degree with ice and snow. Plunging our planet into an ice age that lasted for thousands of years.

The ice slowly melted and carved out huge canyons through out the west and drained into the Pacific Ocean. The Wyoming water drained out through the Green River by first cutting a path through what is now known as Flaming Gorge and Desolation Canyon and then merging with the Colorado River which was draining the western slopes of the Rocky Mountains.

The ice from the most northern part of the impact area most likely drained out by way of the Big Horn Canyon into the Yellowstone River and into the Gulf of Mexico. When the meteorite hit Yellowstone the giant plum of lava and the heat from it all probably created massive flooding of epic proportions and drained out through the Yellowstone River and the resulting volcano's on the west side through out southern Idaho melted the ice sheet's and the water flowed out through the Snake River Plain and also through the Madison River and the Missouri River in southwestern Montana.

As for the great Bonneville Lake and Missoula Lake, when the comet hit it covered the west with ice and also broke the earths crust that caused a lot of the ice to melt by way of many volcanoes sprouting up. Over time with different levels of activity and climate change,   Page 20

The lakes had various shore lines due to the melt water backing up behind ice dams.

The last straw was when the meteorite hit Yellowstone and shattering the ice dams and raising the temperature melting massive amounts of water that flowed down the Snake River Plain and Clark River into the Columbia River the Yellowstone and Madison Rivers creating huge canyons. The Snake River being the major draining point at the time washed a lot of ash and dust that was kicked up by the impact, filling a large area through out the Plain and eventually cutting out through Hells Canyon and into the Columbia River Gorge.

At different times throughout the history of our planet, within the crust there have been large slabs of granite and different types of rock formed that have been pushed up as with the Sierras and the Tetons and I believe that they were pushed up with these great events. As with the Sierras they were lifted due to the compression waves that hit them and the Pacific Plate pushing them up and over the land to the east, the Tetons were fractured by the comet and then pushed almost vertical when the comet hit resulting in there young geologic history.

Page 21

Sleeping Fox Discovered 7.5

Now I would like to talk about Death Valley, the Los Angeles basin and the San Joaquin Valley. As the earth rose and buckled from the east to the west it broke and shoved the Great Sierra Slab of rock into the sky it left a void to be filled that we now call the San Joaquin Valley as I had mentioned before in this book.

On the south side of the valley there is a large curve to the ridge of mountains that isolates the valley from the Los Angeles basin. This was caused when the ground at the end of the great slab broke and the Pacific Plate lurched northward and eastward bending the land in an arc to the north and pushing it up to form the San Gabriel Mountains to the east of the basin forming the high desert to the north where Lancaster, Palmdale and Rosamond sit cradled in the neck of the bend where they may some day be squeezed together.

As for the great Death Valley it is tied in by the same tide of catastrophic sequence of events. Just as with the Grand Canyon, when the ground is pushed with enough force and mass as it raises upward it will crack as it did in Death Valley to enormous depths. When the earth cracked the force that made the great bend in the mountains at the south end of the valley also pulled the ground apart at the point of Death Valley where it is still being torn apart to this day.

As the ice from the comet melted across the west and the ice age that it brought decreased, many inland lakes and sea's came and went washing sediments to the valleys and filling great voids in the earth just as with Death Valley.   Page 22

Page 23

As life and technology expands our understanding of our world, we can look back down the road and see how the people in the past got us here, and so we will always be indebted to our predecessors for there wisdom and getting us to where we are today and weeding out the mistakes. And no matter how much we learn there will always be great voids to fill with the rivers of knowledge that mother earth gives us and hopefully the people in our future can take the technology and knowledge that we have gained and do the same for there understanding.

In this photo you can see the areas around the four corners and within the fox.

In this photo you can see how the rock layers all have the same upward angle across the terrain, proving the sediment that made up this part of the earth was lain down over millions of years, solidifying, and then being lifted in one solid sheet. If it had buckled the strata would be broken and scattered but as it seems there must have been soft areas that have washed away that had not become as solid from being at the bottom of an ancient sea bed.

In conclusion I would like to say that there will be debate's about life and the way thing's come about until the end of time, and I hope if nothing else I have entertained you and stirred up your mind. Once you have seen it, I do know that you will always see the Sleeping Fox and wonder how it got there.

In the world that we live in most religion's look to the sky for there beginning and to the earth for the end, and maybe there is a reason for that, maybe there is a primordial urge to look at the heaven's to see if we can find the past or maybe we know that is where our future come's from.

The End

But let's hope not!   Page 24

Glossary

Ash;   fine mineral particles that form clouds
during a volcano, then fall to the
ground in vast quantities

Asteroid; a large inner solar system rock that
circles between Mars and Jupiter

Astrogeology; the study of space objects such
as near earth objects, meteorites, comets,
and asteroids

Atmosphere; the gaseous area that surrounds
our planet and others

Basin; a bowl shaped area in which water or
other materials can become trapped

Big Horn Mountains; a geographical area
in north western Wyoming

Book Cliffs; a geographical area in eastern
Utah where the earth broke, due to
the impact of the comet, in a vertical
motion resulting in book like features

Caldron; a large pot or bowl shaped unit
with hot bubbling water or steaming
gases coming out

Catastrophic; a disaster of epic proportions

Clark Fork River; a river in western
Montana that flows west into the
Columbia River system

Climatic; as to changing climates, variable
weather patterns of the seasons

Columbia River Basin; the area in the north
Western U.S. that drains the water west of
the divide into the Columbia River

Comet; a large ball of frozen water, gases, and
other materials that orbit's the sun in an
oval pattern and as it nears the sun the
solar wind causes the particles to be
blown off in such a way that it grows a
tail

Composite; being made up of like materials to
form one body

Continent;   any of the great land masses of the
world

Convulsion; an abnormal and violent series of
expansions and contractions

Copernicus; a mathematician and astronomer
from the early 16th century that created
the heliocentric model of our Solar
System

Crater; a hole created by a volcano, meteorite
or comet

Death Valley; an area situated just inside the
the California border and Nevada, the
lowest point in the western hemisphere

Debris; the remains of something broken or
an accumulation of rock fragments

Desolation Canyon; a canyon in which the
Green River flows through in eastern
Utah before joining the Colorado

Dormant; inactive or non functioning

Exhaustion;   depleted of energy

Explosion; the act of exploding

Expulsion; the act of being expelled
or ejected

Extinction; to become extinct or non existent

Fault Line; the area where two pieces of the
earths crust become separated and may
slip and cause earthquakes

Flaming Gorge; a man made lake on the
Wyoming and Utah border named
for a large section of uplifted rock
that appears to be on fire

Four Corners; a point in the southwest part
of the U.S. where the corners of four
state's intersect

Front Range; the east side of the Rocky
Mountains where they meet the Plains
States

Galileo; an Italian mathematician, physicist astronomer and philosopher from the 16th century who was jailed because he believed in the heliocentric model of solar system and now it is believed that he was the father of modern science

Geography; the science of earth structure

Grand Canyon; a large geographic feature of the American southwest

Gulf; a part of the ocean or sea that is mostly surrounded by land

Half Dome; a large rock feature in Yosemite National Park where a glacier cut the rock in half

Heliocentric; the current model of our solar system that put's the sun in the center and the planets revolving around it, see Copernicus and Galileo (16th century)

Hell's Canyon; a deep narrow canyon cut by the Snake River along eastern Oregon and western Idaho borders

Ice Age;   a long period of time when the temperature drops, the climate changes and the earth is covered in snow and ice

Indebted; owing gratitude or recognition to

Inhospitable; non hospitable to life, place's where life is unknown to exist

Isolate; to keep separate from other's

Kaibab Plateau; a large uplifted area on the north rim of the Grand Canyon

Lava; molten rock that flows from a volcano

Lake Bonneville; an ancient lake that covered a large portion of the Great Basin, the big brother of the Great Salt Lake

Los Angeles Basin; a coastal, sediment filled plain

Lake Missoula; a large lake that was formed by ice dam's during the last ice age in north western Montana

Madison River; a head water tributary for the Missouri that start's in Yellowstone

Magma; molten rock within the earths crust

Meteor; a rock that comes into the earth's
atmosphere and burn's up upon entry

Meteorite; a rock that enters earth's
atmosphere and reaches the ground

Missouri River; a river that's head waters
start from the eastern part of the
Continental Divide and eventually
drain into the Gulf of Mexico thru
the Mississippi River

Mogollon Rim; a geological feature that
cuts across upper Arizona and is the
outer perimeter of the lift area that
was caused by the comet

Mount St. Helens; an ancient volcano that
came back to life in 1980 in Washington St.

Nutrient; a nutritional ingredient

Petrified; organic matter that has converted
to stone

Petrified Forest; an ancient forest in
      northern Arizona where the trees have
      turned to stone

Plume; a volcanic plume is a column of hot
      ash and gasses extending from a volcano
      during an eruption

Predecessor; a previous holder of knowledge
      from which we have learned

Prehistoric; a time before written history
      was performed

Pressure wave; the force of pressure radiating
      out from a central point of disturbance,
      it can be air, water or land in earth science

Primordial; from the time of creation

Radiate; to expand from a central point
      outward in a radius

Residual; left over matter after all else
      is gone

Rocky Mountains; a mountain range that runs from northern New Mexico to British Columbia in western U.S. and Canada that divides the flow of our rivers from east to west

San Gabriel Mountains; they form a barrier between Los Angeles Basin and the Mojave desert in California

San Joaquin Valley; the valley runs along the western edge of the Sierra Nevada range in California

Sediment; sand and other materials that settle to the bottom of a liquid after the force of the water becomes to weak to push it on

Shock wave; a large wave of either air or sound pressure that is strong enough to cause a disturbance all at once

Snake River; a river in the western U.S. that
     starts out in western Wyoming and connects
     with the Columbia River and drains into
     the Pacific Ocean

Sierra Nevada's; a mountain range that sit's
     between the San Joaquin Valley in California
     and the Great Basin in Nevada

Solar System; all of our planets with there
     moon's and any celestial bodies that
     orbit our sun

Solidify; the act of changing from liquid
     or gaseous state to a solid

Subduction; the process of one tectonic
     plate being forced under another

Tectonic Plates; the outer skin of the earth
     that floats on a sea of magma and is
     broken into many different sections

Terrain; the lay of the land, the topography

Teton Mountains; a newer mountain range
     that is on the border of Wyoming and Idaho

Theory; a plausible and or scientific explanation of observed facts

Upheaval; to be pushed up from internal forces with great resistance

Vaporize; to change from a solid or liquid state to a gaseous form

Vegetation; the different type's of plant life in a given area

Volcanic; of volcano origins

Waterpocket Fold; a geologic area of eastern Utah that has an almost vertical uplift in a line running north to south that takes in Capital Reef National Park

Yosemite National Park; a geologic area in California that is made up of granite mountains and deep canyons cut and worn by ancient glaciers

Author

Michael J. Branham

www.ingramcontent.com/pod-product-compliance
Lightning Source LLC
Chambersburg PA
CBHW081228170526
45165CB00009B/3001